Introduction to CDMA
2nd Edition

Lawrence Harte
Bruce Bromley
Mike Davis

Phoenix Global Support
2301 Robeson Plaza, Suite 102
Fayetteville, NC 28305 USA
Telephone: +1-910-401-2162
email: MDavis@PGSUP.com
web: www.PGSUP.com

Phoenix Global Support

Copyright 2012 By Phoenix Global Suport
First Printing

Printed and Bound by Lightning Source, TN.

International Standard Book Number: 1-932813-20-9

About the Authors

Mr. Bromley is the Vice President of Training for Phoenix Global Support, a training company who specializes in several modern communications technologies. He has over 25 years of communications experience as a user, trainer, and consultant to multiple government agencies. Mr. Bromley has retired from the U.S. Army after serving over 20 years of active duty in the intelligence community. He has brought his communication knowledge and real-world experiences with him and now uses them to educate the people with a need of this knowledge. He continues to research, learn, and develop new training which keeps Phoenix Global Support ahead of the technology world. Mr. Bromley holds a Bachelors in Information Technology from American Intercontinental University (2005). Mr. Bromley may be contacted at BBromley@pgsup.com.

Mr. Davis, Phoenix Global Support's Chief Operations Officer is also a military veteran with over seventeen years of real-world communications experience for multiple government agencies. He has continued researching and instructing the latest wireless technologies for nearly four years. Today, Mr. Davis maintains multiple relationships with end users of these technologies to ensure that Phoenix Global Support meets each customer's requirements for training and education. Mr Davis may be contacted at MDavis@pgsup.com.

Mr. Harte is the president of Althos, an expert information provider covering the communications industry. He has over 29 years of technology analysis, development, implementation, and business management experience. Mr. Harte continually researches, analyzes, and tests new communication technologies, applications, and services. He has authored over 110 books on telecommunications technologies on topics including Wireless Mobile, Data Communications, VoIP, Broadband, Prepaid Services, and Communications Billing. Mr. Harte holds many degrees and certificates including an Executive MBA from Wake Forest University (1995) and a BSET from the University of the State of New York, (1990). Mr. Harte can be contacted at LHarte@Althos.com.

Phoenix Global Support offers classes in RF propagation, basic/intermediate and advanced GSM theory, UMTS, ground and airborne Geo- Location, tactical integration, WiFi 802.11x theory, Satellite signals intercept (just to name a few), and equipment training for most GOTs and COTs equipment. Through our partnership with an accredited School of business and technology (licensed in North Carolina), we are also able to offer classes on most DOD common communications and networking systems, as well as advanced courses in the Information Technology areas of Microsoft Certified System Administrators/Engineers (MCSA/E), Cisco Certified Network Administrators (CCNA), and Program Management certifications.

Corporate Overview

Phoenix Global Support, LLC. (PGS) offers specialized training on state-of-the art technologies along with a wide range of training services. The company ultimately strives to be a one-stop academy for all training needs, including basic communications, basic/intermediate and advanced geo-location (GL) training, geolocation equipment training and specialized training for the most demanding military, security, or law enforcement mission profiles. PGS is structured to provide education and training through mobile training teams at the customer's location or host on its Fayetteville, NC campus. This flexible structure offers the customer an alternative to higher priced, decentralized training and a wider range of technical training courses.

Table of Contents

Code Division Multiple Access (CDMA)

The Code Division Multiple Access (CDMA) system started as a 2nd generation (2G) digital mobile radio communication system that provided for voice and medium-speed data communication services. Upgrading to CDMA capability from the 1st generation Analog Mobile Phone System (AMPS) allowed CDMA service providers to add more customers for each radio tower (cell site), offer medium-speed data communication services, and provide new high-value information services. As CDMA has improved from 2G to 3G, and now to 4G, its appeal and quality of service, along with increased data rates, has made it the most desired technology for wireless communications in the world.

The CDMA access scheme offers many benefits over the Global System for Mobile communication (GSM) Time Division Multiple Access (TDMA) scheme. So much so, that each 3G technology has implemented CDMA as its air interface access theme. After digitizing the data, it is spread out over the entire available bandwidth. Multiple calls are overlaid on each other on the channel; each one assigned a unique sequence code. CDMA is a form of direct sequence spread spectrum, which simply means the transmitted signal takes up more bandwidth than the information signal that is being modulated.

Another very important benefit that CDMA offers is its capacity, which is much better than a TDMA system. Unlike TDMA, where a single user utilizes an entire timeslot, CDMA is not confined to this. Instead of one person accessing the network during a specified time slot, multiple CDMA subscribers communicate with the network all in the same time slot. With CDMA, time is not an issue. The limiting factor for CDMA capacity is noise. As more subscribers communicate, more noise is added to the RF environment. In an effort to keep the noise down, CDMA has implemented efficient power control algorithms.

Another important benefit of CDMA is quality of service. Due to its unique receivers, CDMA can turn multipath signals into a stronger, better quality signal. These receivers also allow mobiles to perform soft handoffs.

CDMA broadcasts the same cdma_freqs on all towers in the area, which eliminates the need to implement costly frequency re-use patterns.

CDMA History

CDMA has been operational since 1995 and is still going strong. It has evolved from a predominately US and Asia subscriber base into a global presence. Drastic improvements have been made as CDMA has evolved from its original IS-95 configuration (cdmaOne) to 3G form, and then on to cdma2000, followed by Wideband CDMA, and now to 4G Long Term Evolution (LTE).

Here is a brief rundown of the significant highlights in CDMA's history:

1988
CDMA cellular concept

1989
QUALCOMM proposes CDMA as a more efficient, higher-quality wireless technology
CDMA open demonstration conducted in San Diego

1990
First CDMA field trial in New York City

1991
QUALCOMM successfully performs large-scale capacity tests in San Diego

1992

US West orders the first CDMA network equipment
CDMA soft handoff patent granted

1993

CDMA Development Group (CDG) is founded
South Korea adopts CDMA

1994

China performs CDMA field
First Korean CDMA system is unveiled

1995

First commercial launch of cdmaOne (Hutchison Telecom, Hong Kong)
World's first commercial CDMA handset is shipped
CDMA is standardized for U.S. Personal Cellular Service (PCS) band

1996

CDMA is commercially launched in the US (cellular and PCS band), South Korea (cellular band), and Peru

1997

IS-95B standard completed
Canada launches CDMA (cellular and PCS bands)
CDMA is chosen in Japan
First IS-95A WLL (Wireless Local Loop) is launched in India

1998

First CDMA data service is launched in South Korea (IS-95B)
First 1xEV-DO (Evolution Data Optimized) demonstration

1999

CDMA internet services launched in North America, Korea, and Japan

2000
First CDMA2000 1X voice calls completed by Qualcomm, Samsung, and Sprint PCS
CDMA Subscriber Identity Module (SIM) card standard is approved for publication
First CDMA2000 1X data transmission is completed
First CDMA-GSM interoperable SIM card
First CDMA2000 1X service is launched in South Korea

2001
Successful completion of CDMA2000 1xEV-DO trial with Japan
Brazil is first Latin American operator to deploy 3G CDMA2000
Romania launches world's first CDMA2000 network at 450 MHz (CDMA450)
Verizon launches first CDMA2000 1X service

2002
First CDMA2000 1xEV-DO network is launched in US and South Korea

2003
CDMA2000 1X is launched in Mexico, India, and China
CDMA2000 1xEV-DO is launched in Japan

2004
CDMA2000 1xEV-DO Revision A is approved by Third Generation Partnership Project 2 (3GPP2)
Czech Republic launches world's first CDMA2000 1xEV-DO network at 450 MHz (CDMA450)

2005
First demonstration of CDMA2000 1xEV-DO Rev A

2006
CDMA2000 1xEV-DO Revision B standard is published
First CDMA2000 1xEV-DO Rev A network is launched in US

2008

First Advanced Wireless Service (AWS) band is used for CDMA

2009

CDMA2000 1X Advanced standard is published

2010

First EV-DO Rev B network is deployed in Indonesia

2011

First 1X Advanced network is deployed (Cricket Communications)
626.3 million CDMA subscribers worldwide

CDMA Deployment

As of June 2012, there were 364 CDMA operators worldwide in 124 different countries for a total of 685 CDMA networks, providing service to 626.3 million subscribers worldwide. Of the 364 operators, there were 120 CDMA 2000 operators in 66 countries. There were 115 CDMA operators in 60 countries using the 450 band. There were 169 CDMA equipment manufactures producing 3170 types of equipment. Data from CDG.com on 7/9/2012

CDMA Evolution

The original CDMA standard was called cdmaOne, also known as IS-95A and IS-95B (IS stands for Interim Standard). Prior to this technology, AMPS was the cellular standard in America. CDMA offered 10-12 times more capacity than AMPS, and 4-5 times more capacity than GSM. IS-95B was introduced a few years after IS-95A, and added 14.4kbps circuit switched data capability, and up to 64kbps packet switched data capability.

CDMA2000 1X began in 2000 with IS-2000. As with any newer cellular technology, the biggest improvements were the data capabilities. CDMA2000 introduced 144kbps packet data capability. CDMA2000 also made modifications and improvements to increase capacity and efficiencies. CDMA2000 supports 1.5 to 2 times more voice channels than IS-95. CDMA2000 is 21 times more efficient than AMPS and 4 times more efficient than TDMA networks. The 1X indicates that it is using one 1.25 MHz spread channel. Before the signal is spread to 1.25 MHz, in 850 MHz the original channel is 30 kHz, and in the 1900 band the signal starts as a 50 kHz channel.

When cdmaOne became cdma2000, few enhancements were made to improve the system and increase productivity and capacity. The battery life was improved by adding new power controls to the BTS. Additionally, new signaling channels reduced the amount of time the handset was active on the network. New Walsh codes were added to increase user capacity. New data and signaling channels were added to increase efficiency while increasing throughput. More radio configurations were added, creating more flexibility for the network when allocating radio resources, bandwidth and speed. The IS-2000 standard was adopted and implemented through the world.

CDMA2000 1X EV-DO (EVolution Data Optimized) (IS-856) really improved the data rates by adding additional channels that for use with data traffic only. It also introduced a Packet Switched Core Network to its architecture. CDMA2000 1X EV-DO utilized an adaptive modulation known as 16-QAM (16th Quadrature Amplitude Modulation). Additionally, CDMA2000 1X EV-DO used Time Division Multiplexing (TDM) for users. Most network operators implemented both 1X and EV-DO in their systems. Though the two systems are collocated, they are separate entities due to the different protocols (IS-2000 and IS-856). This resulted in subscribers having the ability to either be in a voice call or a data session, but not both simultaneously. The identifying information for a CDMA2000 1X network provider and a CDMA2000 1X EV-DO provider is different, despite being from the same provider. Release 0 offered 2.4 Mbps downlink speeds and 153 kbps uplink speeds, while release A offered a 3.1 Mbps downlink and a 1.8 Mbps uplink.

CDMA2000 3X (also known as EV-DO revision B or Multi-carrier) can implement three 1X channels, which resulted in an ultimate bandwidth of 5 MHz. This bandwidth is required as a 3G standard and is known as Wideband, as in WCDMA (Universal Mobile Telecommunications System -

UMTS). CDMA2000 3X had not been implemented throughout the world yet. CDMA2000 3X offered packet data speeds of up to 9.6 Mbps on the downlink and up to 5.4Mbps on the uplink. The voice capacity was increased 1.5 times in CDMA2000 3X.

CDMA2000 1X EV-DV (also known as 1X Advanced) was implemented in December 2011 by Cricket Wireless. It allowed the user to simultaneously talk, send and receive simultaneously on the same channel. This eliminated the need for a second 1.25 MHz channel, and allowed for an increase in voice capacity of 4 times that of previous versions. The packet data speeds were significantly higher with rates of up to 32 Mbps on the downlink and up to 12.4 Mbps on the uplink.

As the CDMA protocols were revised, associated numbers were assigned to reflect the changes made. There are seven main protocol revisions (P_REV). P_REV 1 is the original revision used with IS-95 in the 1900 band. P_REV 2 was termed IS-95A and used only in the 1900 (PCS) band. P_REV 3, known as Technical Services Bulletin 74 (TSB-74) was an improvement on the IS-95A protocol. P_REV 4 is IS-95B phase 1, and it included interoperability in both the 850 and 1900 bands. P_REV 5 is IS-95B phase 2. P_REV 6 includes the cdma2000 1X and EV-DO Release 0. P_REV 7, is 1X EV-DO Revision A.

Network Architecture

The components that make up CDMA network architecture are very similar to a GSM network, with a few exceptions. Their functions are also extremely similar. The components that make up the CDMA network fall within three main categories, which include the Circuit Switched Core Network (CSCN), the Packet Switched Core Network (PSCN), and the Radio Access Network (RAN).

Figure 1.1 shows that the Circuit Switched Core Network (CSCN), which has been around since CDMA was first introduced as a 2G technology, handles all of the circuit based switching for calls, subscriber services and the low rate data, such as Short Message Service (SMS). The Radio Access Network (RAN) only consists of two components. These components spend all of their time dealing directly with all of the subscribers under their control. The Mobile Station (MS) is simply the phone. It is the reason for the entire network and why the entire infrastructure exists.

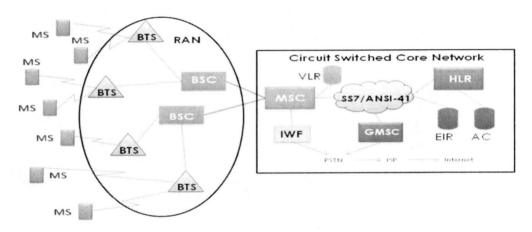

Figure 1.1, CDMA Architecture

Circuit Switched Core Network (CSCN)

The first component in the CSCN is the Home Location Register (HLR). The HLR is a subscriber database containing all of the subscriber's information for each subscription, both pre-paid and post paid. Every MS is assigned to a single HLR, which means that there is one logical HLR per network with multiple physical locations. The location of the HLR will be based on where the subscriber signed up for service. There are multiple HLR's to help with faster access and information storage and back up. These HLRs hold the same data, which can be either permanent or temporary.

The permanent data is held for every subscriber who belongs to that network. Some of the data consists of what services they are allowed to access (what they are paying for), roaming agreements between networks, and all of the identifiers, both subscriber and equipment. Some of these IDs include the Electronic Serial Number (ESN), Mobile Equipment Identifier (MEID), International Mobile Subscriber Identity (IMSI), Mobile Identity Number (MIN), and Mobile Directory Number (MDN). These identifiers will be discussed later. As far as temporary data goes, in order for the network to properly route incoming calls and data, it needs to know where the subscriber is currently being serviced. This area of service falls under the SID/NID/Reg_Zone location. This identifier is similar to a Location Area Identifier (LAI) in GSM.

Another component of the CSCN, which is typically co-located at the HLR, is the Authentication Center (AC). Like the AuC in GSM, it is part of the process in Mobile Station (MS) authentication. Every MS holds a key called the A-Key, which is unique to each MS, and is also held at the AC. The AC begins the authentication process by combining the A-Key, Electronic Serial Number (ESN), and Random Number (RAND) through an algorithm called CAVE (Cellular Authentication Voice Encryption). This produces two things; an SSD_A (Shared Secret Data_A), which is used for authentication, and an SSD_B (Shared Secret Data_B), which can be used for voice privacy.

A registry called the Equipment Identity Register (EIR) is another component of the CSCN used to validate the actual equipment used in an effort to mitigate piracy and cloning. The equipment is now identified by its Mobile Equipment ID (MEID) number. The MS's MEID is compared against three databases located here, which include the Normal, Block and Track lists. The Normal list is simply a list of valid MEID ranges. The Block lists are MEIDs that should be denied service due to either being reported stolen or having service-impacting hardware issues. The Track lists are MEIDs that have been reported lost or have minor hardware issues. As in GSM, the EIR is an optional database that the networks may use.

The Mobile Switching Center (MSC) is the link from the RAN into the CSCN. The MSC is a sophisticated telephone exchange that provides circuit switched calling, mobility management and CDMA services in a geographical area of coverage. It is responsible for activities like registration, location updates, paging, call setup and call teardown. The MSC controls numerous Base Station Controllers (BSCs) within its area of responsibility.

The Visitor Location Register (VLR) component of the CSCN is a database that holds information on each MS that is currently registered within the MSC controlled area. It is a temporary database which holds both permanent and temporary MS data. It is temporary because when an MS leaves the area, the data is typically erased. The temporary storage of the MS's permanent data includes the IMSI, MEID, MIN, service profile and HLR address. The temporary data stored there includes TMSI, if used, current and previously visited SID/NID/REG_ZONE, authentication data, authorization data and the Temporary Local Directory Number (TLDN).

The next component of the CSCN is the Gateway MSC (GMSC). The GMSC is a fully functional MSC that has links outside of the local network. Some of these links include; PSTN (Public Switched Telephone Network), ISDN (Integrated Services Digital Network), CSPDN (Circuit Switched Public Data Network), PSPDN (Packet Switched Public Data Network), other Public Land Mobile Networks (PLMN), and other networks. When a subscriber requires these services, they are routed to the appropriate GMSC to gain access outside of the home network.

The last component in the CSCN is the Inter Working Function (IWF). The IWF is primarily a 2G component that allows the user to get on the Internet. Its main function is to convert data to and from the MS and the Internet. It converts a PCM (pulse code modulation) data stream that the MSC uses into modem tones that the PSTN understands. This technique is similar to good old dial-up.

Radio Access Network (RAN)

Figure 1.2 shows that the RAN is the work horse of the network. The elements which make it up deal directly with all of the Mobile Stations within its control. It consists of two components; the Base Station Controller (BSC) and Base Transceiver Station (BTS). The BSC's responsibilities include voice routing and circuit switched data routing from the MS into the CSCN.

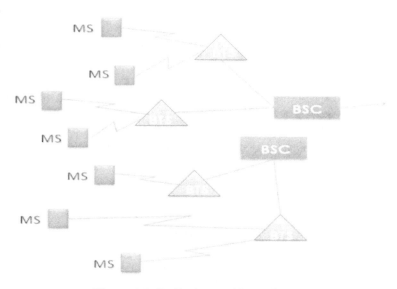

Figure 1.2, Radio Access Network

The BSC conducts some power control functions for the multiple base transceiver stations it supports. It also deals with mobility management as it controls and directs handoffs from one cell site to another.

The BTS provides the air interface and is the direct link between all MSs and the network. It manages RF resources such as frequency assignments, sector separation and power control.

BTSs, whether they are GSM, UMTS or CDMA, all have the same basic appearance and type. Large Macro towers are primarily found in rural environments where the subscriber density is fairly low. These sites are located on high towers, are spread out over large distances, and transmit at higher wattages.

In more densely populated areas, the towers and coverage become smaller. A small Macro tower with a coverage area of approximately 1-5 KM is typically found in suburban areas.

In urban areas, where the subscriber density is extremely high, Micro cells are used. These antennas are typically mounted on structures already in place. This is known as parasitic. There are many of these structures with coverage of only 100M - 1 KM.

Pico cells are typically used in areas where Macro and Micro tower signals can't penetrate due to structure or building height or in structures with extremely high subscriber density. Pico cells are frequently found in subways, airports and sports arenas, and their coverage is only 20M-100M.

A newer type of cell tower is the Femto cell. These are used by both businesses and residences. Businesses use Femto towers to fill the gaps caused by all of the other types of cells. Typically, an office building will use a combination of Pico and Femto cells to provide adequate coverage. In neighborhoods with minimal or no coverage, a person can purchase a Femto cell, which will provide cellular coverage around the residence. Femto cells are directly connected to the Internet, which is how information is sent and received from the network. The Femto cell antenna can cover an area from 20 - 50 meters.

Several specialized cells are also available, including the umbrella cell and the Cell On Wheels (COW). An umbrella cell is typically a large macro tower located in a dense environment used to help with the capacity load during high activity times. A COW is a temporary, mobile tower used to supply or add to existing service. These can be used during natural disasters when normal service is non-existent, or to aid with a sudden capacity issue. They are also used at special events where normal subscriber density would be very low.

All cellular network towers can be sectored. Typically, they are either Omni-direction (non-sectored), Bi-sectored, or tri-sectored. Sectoring increases capacity of the base station. In CDMA, sectors are identified by their cdma_freq, PN_offset (pseudo noise offset), and Base Station ID (BSID). Omni towers are typically Pico cells and Femto cells. Bi-sectored towers are usually found along highways and railways. The most common are tri-sectored towers, which provide coverage all around the tower.

Packet Switched Core Network (PSCN)

Figure 1.3 shows that the Packet Control Function (PCF) actually resides in the RAN with the BSC. Its job is to route Internet Protocol (IP) packet data from the mobile stations to the Packet Data Serving Node (PDSN) located in the Packet Switched Core Network. It also handles the data buffering and rate conversions needed. Based on need, it assigns the Supplemental Channels (SCH) required to send and receive the packet data. This figure shows that the Packet Data Serving Node (PDSN) manages the radio-packet interface between the RAN and the IP network. When an MS needs to connect to the IP network, it is the PDSN who establishes a connection using Point-To-Point Protocol (PPP). This protocol is used to establish a temporary connection on to an IP network. The PDSN also collects usage data sent from the AAA (Authentication, Authorization, and Accounting) server. The PDSN can support both simple and mobile IP.

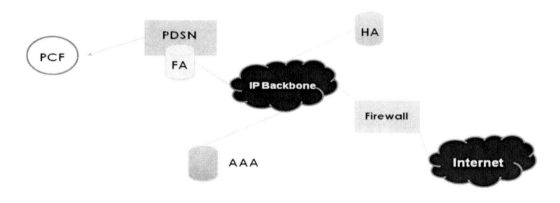

Figure 1.3, Packet Switched Core Network

Simple IP means that an MS establishes an internet connection and is issued a temporary IP address. The MS does not leave a PDSN controlled area and is still controlled by the issuing PDSN. The key is that the originally issued IP address never changes.

When an MS is mobile and connected to the Internet, it may move into another PDSN-controlled area. This usually means that it will receive a new IP address. Instead of having the connection terminated and re-established, it is maintained. This ability to maintain the connection and change the controlling PDSN is known as Mobile IP.

The Foreign Agent (FA) has a few roles. It works and is co-located with the PDSN. It is the component which issues the temporary IP address. This is known as a Care-Of-Address (COA). It stores information about each MS that visits the PDSN area. When an MS is in a data session and crosses into its PDSN area, the new IP address and location will be forwarded to the Home Agent (HA) to route and maintain the internet connection.

The AAA (Authentication, Authorization, and Accounting) works for the PDSN to authenticate and authorize access. It collects and stores data for billing and invoicing purposes. Remote Access Dial-In User Service (RADIUS) or DIAMETER is the protocol used to access the AAA server.

The last component is the Home Agent (HA). It is basically a router that works with the FA to provide the Mobile IP functionality. It tracks the location and IP address of the MS as it moves from one PDSN to another. It routes the data to the mobiles that are currently attached to a foreign network.

Mobile Station Identifiers

There are several ways to identify the Mobile Station (MS) in CDMA, such as by using the Electronic Serial Number (ESN). The primary way of identifying the MS today is by the Mobile Equipment Identifier (MEID). Thanks to backwards compatibility, an MS can also be identified by using the pseudo Electronic Serial Number (pESN).

To identify the subscriber in CDMA there are multiple identifiers that can be used, such as the Mobile Identity Number (MIN), the Mobile Directory or Dialed Number (MDN), the International Mobile Subscriber Identity (IMSI), and the Removable User Identity Module (R-UIM). All of these numbers are used at different times within the network.

Mobile Identification Number (MIN)

The Mobile Identifier Number (MIN) is a ten digit number that is assigned to a mobile station to identify the subscriber. It is divided into MIN1 and MIN2. MIN1 is the 7 digit portion of the number. MIN2 is the 3 digit area code portion of the number. It was originally used as the subscribers' actual phone number, but we now have the ability to port our telephone numbers between regions and between providers, so the MIN is not always the phone number of the subscriber.

Mobile Directory Number (MDN)

A Mobile Directory Number (MDN) is a number that can be dialed through the public telephone network. A full mobile directory number contains the country code, national designation code and subscriber number, and it can be up to 15 digits. Modern Handsets can have the ability to maintain more than one MDN in the hardware.

Electronic Serial Number (ESN)

The original way to identify the actual handset was the Electronic Serial Number (ESN). ESNs are transmitted from the mobile device to the system when an attempt is made to to access (get service from) the system. This is an eight digit hexadecimal number that is comprised of the two digit manufacturer code and a six digit serial number. ESN's were carried over from IS-95 into CDMA 2000 and, in June of 2010, they were exhausted. They have since been replaced with Mobile Equipment Identifiers (MEID) and Pseudo Electronic Serial Numbers (pESN) for backwards compatibility between CDMA2000 and IS-95. To find the ESN, look underneath the battery of a device where it will be printed on the sticker. Some handset models will also display the ESN within the handset settings.

Mobile Equipment Identity (MEID)

The current method of identifying the equipment (handset/terminal) is by using the Mobile Equipment Identifier (MEID). This is a globally unique number identifying a specific MS. It is similar to the International Mobile Equipment Identifier (IMEI) used in GSM, except it is in hexadecimal format. The MEID is 15 hexadecimal digits in length. It is broken down into a two digit regional code followed by a six digit manufacturer code, which is then followed by a six digit serial number and a single digit check digit. This number can be found printed on the underside of all CDMA handsets today. The check digit uses the Luhn formula to attempt to prevent cloning of the handset. The Luhn formula is a simple algorithm that combines the first fourteen digits to produce a single value for the 15th digit.

psuedo Electronic Serial Number (pESN)

To maintain backwards compatibility between CDMA2000 and IS-95, the system uses pseudo Electronic Serial Numbers (pESN). The 15 hexadecimal digit MEID is replaced with an eight digit pESN. This is accomplished using the Secure Hash Algorithm -1 (SHA-1) to reduce the 15 digit MEID into the eight digit pESN. The use of a pESN is required to generate the Public Long Code Mask (PLCM), which is used to uniquely identify each user on the network. pESN's are easily identified by the manufacturer code of 80 in hexadecimal format. Usually seen as 0x80, this number converted into a decimal is 128. This manufacturer code is annotated as "reserved for pESN's."

International Mobile Subscriber Identity (IMSI)

The International Mobile Subscriber Identity (IMSI) is utilized in CDMA 2000. Just as in GSM, the IMSI is typically 15 digits. It is broken into three sections, which include the Mobile Country Code (MCC), consisting of three digits, the Mobile Network Code (MNC), consisting of two digits, and the Mobile Subscriber Identification Number (MSIN) or Mobile Identification Number (MIN). The MNC is also identified as the IMSI_11_12. This simply refers to the 11th and 12th digits from the right of the IMSI (ex. 310/01/9876543210). Counting from right to left, the 11th number is 1 and the 12th number is 0. This is the MNC 01.

There are two IMSI classes. A Class 0 IMSI contains 15 digits, while a Class 1 IMSI contains less than 15 digits. The most common class is Class 0. The use of IMSI's by CDMA providers allows for interoperability between providers and between technologies, such as GSM.

There are several types of IMSI's used in CDMA networks. The type being used by a particular provider will be displayed via the system parameters message, which is broadcasted on the Forward - Broadcast Control Channel (F-BCCH).

The IMSI_T or True IMSI is the actual 15 digit number that consists of the MCC/MNC (IMSI 11_12) and Mobile Subscriber Identification Number (MSIN). This is the same type of IMSI used by GSM and UMTS. The MSIN is not the same 10 digit number as the Mobile Identification Number (MIN). This IMSI type is generally not used by the networks.

The IMSI_M is a 15 digit number that consists of the MCC/MNC and the MIN. The Mobile Identification Number (MIN) replaces the MSIN (Mobile Subscriber Identification Number). The MIN is not the same number as the MSIN.

The IMSI_S or IMSI Short is just the 10 digit Mobile Identification Number (MIN) without the MCC or MNC. In most cases, this is the preferred type of IMSI used by providers.

The IMSI_O or IMSI Operational is not an actual type of IMSI; it is just what type of IMSI the network is currently using. i.e. IMSI_M (sent of the F-BCCH).

The Mobile Station Identification (MSID) is used by the network to offer guidance to mobile stations as to how they will identify themselves. MSID_type 0 tells the MS that it will identify itself by IMSI_S and ESN (used in Band Class 0 only). MSID_type 1 tells the MS that it will identify itself by ESN only. MSID_type 2 tells the MS that it will identify itself by its IMSI only. MSID_type 3 tells the MS that it will identify itself by its IMSI and ESN (or pESN). MSID_type 4 tells the MS that it will identify itself by TMSI. 3GPP2 C.S0004-A (pg 2-22)

Temporary Mobile Subscriber Identity (TMSI)

A Temporary Mobile Station Identity (TMSI) is a 32-bit (8 hexadecimal digits) number that is used to temporarily identify a mobile device that is operating in a local system. A TMSI is typically assigned to a mobile device by the system during initial registration. The TMSI is used instead of the International Mobile Subscriber Identity (IMSI) or the mobile directory number (MDN). TMSIs may be used to provide increased privacy (keeping the permanent identifiers private), and to reduce the number of bits that are sent on the paging channel (the number of bits for a TMSI are much lower than the number of bits that represent an IMSI or MDN).

IP Address

Internet Protocol addressing (IP addressing) is the use of unique identifiers in a data packet, which are assigned to a particular device or portion of a device (such as a port) within a system or a domain (portion of a system). IP addressing varies based on the version of Internet protocol. For IP version 4 (IPv4), this is a 32-bit address, whereas for IP version 6 (IPv6), this is a 128 bit address. To help simplify the presentation of IPv4 addresses, it is common to group each 8 bit part of the IP address as a decimal number separated from other parts by a dot (.), such as: 207.169.222.45. For IPv6 it is customary to represent the address as eight, four digit hexadecimal numbers separated by colons, such as 1234:5678:9000:0D0D:0000:5678:9ABC:8777.

While the CDMA system was not designed to directly use IP addressing, an IP address can be assigned to mobile devices when they are accessing data networks (such as the Internet).

The CDMA system permits the static or dynamic assignment of IP addresses. Static IP addressing can simplify the connection of services to mobile devices. Dynamic IP addressing can better manage a limited number of IP addresses and enhance the security of systems.

Static IP Addressing

Static IP addressing is the process of assigning a fixed Internet Protocol (IP) address to a computer or data network device. Use of static IP address allows other computers to initiate data transmission (such as a video conference call) to a specific recipient.

Dynamic IP Addressing

Dynamic IP addressing is a process of assigning an Internet Protocol address to a client (usually an end user's computer) on an as-needed basis. Dynamic addressing is used to conserve on the number of IP addresses required by a server, and to provide an enhanced level of security (no pre-defined address can be used by hackers).

Removable User Identity Module (RUIM)

The Removable User Identity Module (R-UIM) is the CDMA2000 equivalent to a Subscriber Identity Module (SIM) in GSM. The R-UIM contains, at a minimum, the IMSI, MDN, Authentication Key (A-Key), Shared Secret Data (SSD) A/B, and Preferred Roaming List (PRL).

The use of the R-UIM is optional for CDMA 2000. However, if R-UIM's are used, the IMSI associated with that particular MS will be changed to the IMSI of the R-UIM. Additionally, R-UIM may change the ESN or MEID of the MS as the R-UIM has its own User Identity Module Identification Number (UIMID), or Enhanced Used Identity Module Identification Number (E-UIMID). Any handset that is not capable of using an R-UIM is called Mobile Equipment (ME).

Location Identification

A System Identifier (SID) is a number (1-5 digits) that is used to signify a location of service associated with a CDMA network provider. Typically, an SID covers an area larger than the size of a city. It can cover more than that in areas of lower populations. For example SID 21 is Verizon Wireless for the Ohio cities of Sandusky, Cleveland, Toledo, Columbus, Dayton, and Cincinnati, as well as the Michigan cities of Detroit, Ann Arbor, Flint, Lansing, East Lansing, Grand Rapids and Saginaw-Bay.

A Network Identifier (NID) is used to identify a location of service within a given SID. An NID is always associated with a SID; therefore it is always identified as a SID/NID pair. Depending on capacity issues, the network may use multiple NIDs. Other places might include only 1 NID in a SID.

Another method to control capacity issues is the implementation of registration zones (Reg_Zone) within the SID/NID. These Reg_Zones are similar to Location Area Codes (LAC), which are used in both GSM and UMTS networks, as far as location updates are concerned.

Preferred Roaming List (PRL)

The Preferred Roaming List (PRL) resides on either the R-UIM or the ME. This list is what enables the MS to determine if it can roam outside of its home network. Additionally, information is stored on this list that is used during the system selection and acquisition process. The PRL includes:

* Prioritized Bands / Sub-bands

* Prioritized SID/NID Pairs

* Preferred Channel Sets

* May include: Subscriber User Zones w/ allowable Base Station IDs

Number Portability

Number porting is when a subscriber moves the phone number to a geographic region outside of that within which it was originally purchased, or switches to a different provider, but keeps the same number that was originally assigned by the former provider. Typically, the MIN and MDN are not the same number. Users who port their numbers will be assigned a new MIN, but will keep the original MDN.

Number portability allows customers to change service providers without having to change telephone numbers. Number portability involves three key elements: local number portability, service portability and geographic portability. To enable number portability, the CDMA system maintains a number portability database (NPDB). This database helps to route calls to their destinations, which may have assigned telephone numbers that are different (number has been ported) from the destination phone numbers.

MS Power Classes

MS Power Classes are designated by band and have a minimum transmit power and a maximum transmit power. Class 1's Effective Radiated Power (ERP) must be between 1.25 watts and 6.3 watts. Class 2's Effective Radiated Power (ERP) must be between .5 watts and 2.5 watts. Class 3's Effective Radiated Power (ERP) must be between .2 watts and 1 watt. The MS power classes are specific to different CDMA bands. For example, cdma_450 would use Class 1, cdma_850 would use Class 2, and cdma_1900 would use Class 3.

Wireless Local Loops (WLL)

Wireless Local Loop (WLL) is the successor to copper local loop, which uses an analog connection for in-band signaling and digital connection for out of band signaling (ISDN). It is ideal for areas with remote access and areas that do not have any communication infrastructure in place. A WLL terminal generally consists of a handset connected to a unit. The terminal may have multiple connectors for access to a computer or fax machine.

WLL networks can include more affordable infrastructure, operations and maintenance costs. Also, WLL terminals may be setup to prevent mobility. They can be network or area dependent, or may lock to a particular sector, BaseID, NID, Reg_Zone, or User Zone. WLL networks do not have heavy mobility management loads to carry, which means cheaper operation, and they can also be configured to provide high speed Internet.

One of the characteristics that make WLL unique is the fact that it is geographically specific. Even though it is based on CDMA's wireless network, the terminals are not all mobile. WLL offers various services and are all subscription based. User zones may be supported by a public system on the same frequency as the base station, or on a private system operating on a different frequency.

There are two types of User Zones in a wireless local loop; Broadcast User Zones and Mobile Specific User Zones. The Broadcast User Zone is a list of available user zones that are within a particular BTS Coverage area. The MS maintains a list of allowed user zones that prevent mobility. This list is sent as part of the BCCH message known as the User Zone Identification Message. This allows for the use of fixed terminals within private residences or Public Call Offices. The User Zone Identification Message is compared to what is pre-programmed within the MS to determine if the BTS is available for use by the MS.

Mobile Specific User Zones are not broadcasted by the network, but rather the MS uses some of the overhead messaging broadcasted by the BCCH to determine if the MS may register and use a particular sector. The MS receives the broadcasted information and compares it to what is pre-programmed in the MS to determine network availability. Some of the parameters can include the SID, NID, Base ID, and LAT/LON. This allows for the MS to be fixed to a particular BTS, or a small group, within an immediate area.

CDMA Characteristics

The CDMA interface is very different than the standard TDMA interface. The utilization of codes, spreading, and channel usage, makes CDMA unique. This chapter will discuss the characteristics of CDMA.

Direct Sequence Spread Spectrum

Spread spectrum technology is the ability to manipulate small user data streams over a much larger bandwidth. CDMA uses Direct Sequence Spread Spectrum (DSSS) which spreads the 30 kHz or 50 kHz (850 Band for 30 kHz and 1900 Band for 50 kHz) data signal over a 1.25 MHz bandwidth. The use of direct sequence means that the same pattern is used for spreading the 30kHz or 50kHz of information over the entire 1.25 MHz bandwidth. This helps CDMA by reducing the interference and jamming of the desired signal by spreading it across the noise floor. Keeping the signal in the noise floor allows for multiple users to interact with the network on the same frequency at the same time. Each user has a specific code, allowing multiple users to occupy the same frequency at the same time. The spreading is accomplished by means of a Walsh Code that is independent of actual user data.

cdma_freq

In CDMA, physical channels are comprised of a Forward and Reverse Frequency. A physical channel is the actual RF transmission path between the MS and the BTS.This pair of frequencies is known as a cdma_freq. The independent data streams are also known as Downlink and Uplink. Forward/Downlink is the transmission frequency coming from the BTS (tower) to the MS. Reverse/Uplink is the transmission frequency from the MS to the BTS. The cdma_freq is a number that represents the actual frequency pair, (e.g., cdma_freq 384 = Downlink frequency of 881.520 and an Uplink frequency 836.520).

These cdma_freqs are divided into pre-set ranges known as bands. These bands are further divided into sub-bands. Wireless providers lease the sub-bands that they require to provide service to their subscribers. In CDMA, the providers will typically require less spectrum than in a TDMA based network. This is because there is no need for frequency reuse, as each BTS in a given area will broadcast the same cdma_freq.

There is always a buffer or offset between the Forward and Reverse frequencies. This offset varies between the bands. In the most widely used bands, the higher the band, the higher the offset between the forward and reverse frequencies.

Frequency Reuse

Frequency reuse is the process of using the same radio frequencies (cdma_freq) on radio transmitter sites within a geographic area that are separated by sufficient distance to cause minimal interference with each other. Frequency reuse allows for a dramatic increase in the number of customers that can be served (capacity) within a geographic area on a limited amount of radio spectrum (limited number of radio channels). The ability to reuse frequencies depends on various factors, including the ability of channels to operate with interference signal energy attenuation between the transmitters.

The CDMA radio channels use coded channels that are uniquely assigned to each user. This allows many users to operate on the same frequency. This also allows frequencies to be reused in every cell site and sectors within a cell site. However, the use of the same frequency in the same cell site and sector increases the interference levels and decreases the capacity of the radio channels.

Figure 1.4 shows how CDMA systems can reuse the same frequency in each cell site. This example shows that the frequency use factor is 1 (N=1), and that the overlap of the radio channels results in an increased interference level in the overlapping area. Because multiple chips represent each channel, this overlap simply results in the loss of some of the chips and this reduces the capacity of the CDMA system.

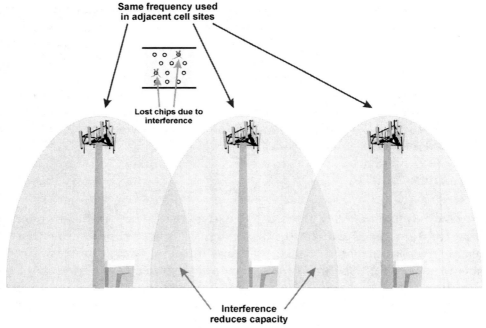

Figure 1.4 , CDMA Frequency Reuse

Using the same frequency in each cell site within a cellular system eliminates the need for frequency planning. However, to reduce the interference, different code sequences may be used in each cell site. This is called PN offset planning. The CDMA system defines 512 PN offsets that are spaced 64 chips apart in time. The time period for 64 chips equates to approximately 10 miles of signal coverage

Multi-Carrier

Multi-Carrier (MC) in CDMA is when a provider deploys more than one 1.25 MHz cdma_freq per sector. This is done to increase capacity while continuing to provide high quality service. Most network providers employ some sort of multicarrier capability throughout their networks. An example would be the broadcasting of cdma_freq 25,100, and 150 on the same sector of a BTS.

Figure 1.5 shows an additional Cdma_freq to the tower. It doubles the capacity and each channel works independent of the others. Subscribers are assigned to one of the other Cdma_freqs based on Hashing.

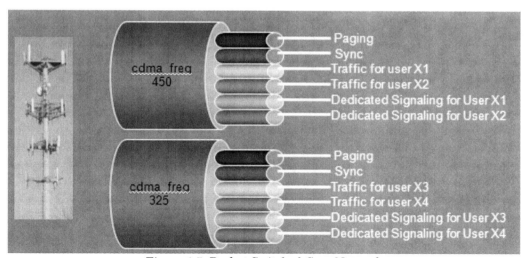

Figure 1.5, Packet Switched Core Network

CDMA Bands

There are 19 CDMA band classes according to the 3GPP specifications. (GPP2 C.S0057-C v1.0) The most widely used CDMA Bands are Band Class 0, Band Class 1, Band Class 5 and Band Class 15.

Band Class 0, also known as the Cellular 850 band, has channels numbered from 1 - 1023. The frequency range for band class 0 is 869MHz - 894MHz for the Downlink/Forward and 824MHz - 849MHz for the Uplink/Reverse. The channel offset (Buffering of the Duplex channel) between the forward and reverse Cellular 850 band is 45MHz of spacing. Since DSSS is used in CDMA, the typical channel step size is 41. This means that, although there are channels available from 1-1023, for every channel utilized, the next available channel will be 41 channels later. This step size can vary between providers.

Band Class 1 is also known as the PCS 1900 band, and has channels numbered from 0 - 1200. The frequency range for the PCS 1900 band is 1930MHz - 1990MHz for the Downlink/Forward and 1850MHz - 1910MHz for the Uplink/Reverse. The channel offset between the forward and reverse PCS 1900 band is 80MHz. The channel step size 25.

Band Class 5 is known as the 450MHz band, and has channels numbered from 564 - 1985. The frequency range for the 450MHz band is 421.675MHz 493.48MHz for the Downlink/Forward and 411.67MHz - 483.48MHz for the Uplink/Reverse. The channel offset for band class 5 is 10MHz. The channel step size is 50.

Band Class 15 is known as the Advanced Wireless Service (AWS) 1700MHz band, and has channels 0 - 899. The frequency range for the AWS band is 2110MHz - 2155MHz for the forward/Downlink and 1710MHz - 1755MHz for the Reverse/Uplink. The channel offset for band class 15 is 400MHz. The channel step size is 25.

Channel Structure

Channel structure is the division and coordination of a communication channel (information transfer) into logical channels, frames (groups) of data, and fields within the frames that hold specific types of information.

Each CDMA communication channel is composed of radio channel pairs. Each coded communication channel (traffic channel) is divided into 20 msec frames, and each frame is divided into 1.25 msec time slots (power control groups). The duplex channel spacing varies based on the frequency band. The 20 msec frames for the forward and reverse channels may be transmitted with a time offset relative to each other. The ability to change the time offset can be used to coordinate the transmission of power control groups.

Modulation

Modulation is the process of changing the amplitude, frequency or phase of a radio frequency carrier signal (a carrier) to change with the information signal (such as voice or data). Modulation efficiency is a measure of how much information can be transferred onto a carrier signal. In general, more efficient modulation processes require smaller changes in the characteristics of a carrier signal (amplitude, frequency or phase) to represent the information signal. The CDMAone and CDMA2000 radio channel uses multiple types of Modulation including BPSK, QPSK, 8-PSK and 16-QAM.

Binary Phase Shift Keying (BPSK)

Binary phase shift keying (BPSK) is a modulation process that converts bits into phase shifts of the radio carrier without substantially changing the frequency of the carrier waveform. The phase of a carrier is the relative time of the peaks and valleys of the sine wave relative to the time of an un-modulated "clock" sine wave of the same frequency. BPSK uses only two phase angles, corresponding to a phase shift of zero or a half cycle (that is, zero or 180 degrees of angle).

Quadrature Phase Shift Keying (QPSK)

Quadrature phase shift keying (QPSK) is a type of modulation that uses 4 different phase shifts of a radio carrier signal to represent the digital information signal. These shifts are typically +/- 90 and +/- 180 degrees.

The CDMA system uses different types of quadrature modulation on the forward and reverse channels. The forward channel uses quadrature phase shift keying (QPSK) and the reverse channel uses offset quadrature phase shift keying (O-QPSK).

The difference between QPSK and O-QPSK is that offset quadrature phase shift keying requires that the transmitter does not pass through the origin point (0 signal level) when it is changing signal phases. By using O-PSK, the RF amplifier does not have to be as linear (precise) and this allows the use of a less expensive and more efficient (longer battery life) transmitter.

8-Phase Shift Keying (8PSK)

8-Phase Shift Keying (8PSK) is a type of modulation that uses 8 different phase shifts of a radio carrier signal to represent the digital information signal.

16 Quadrature Amplitude Modulation (16-QAM)

16 Quadrature Amplitude Modulation (16-QAM) is a form of amplitude modulation. The amplitude of two waves, 90 degrees out-of-phase with each other is modulated to represent the data signal. Amplitude modulating two carriers in quadrature can be viewed as both amplitude modulating and phase modulating a single carrier. The percent of amplitude and the degree of phase result in one of sixteen different binary strings.

Power Control

Power Control in CDMA is critical. Too little power and the BTS can't hear the MS. Too much power and no one can hear anything because power equals noise. If there is too much noise, the number of users will be decreased causing great inefficiency in the system. There are power controls on both the forward and reverse links.

Forward power control means that the power coming from the forward/downlink is power controlled. The two types of forward power control are known as open loop and closed loop.

Open loop power control is adjusted while the mobiles are in idle mode (not in a call or data session). The BTS will continually reduce its power. Its goal is to reduce its power as much as possible (remember; power equals noise). The mobiles report how they hear the BTS based on the Frame Error Rate (FER). The BTS will increase its power to correct for this. In its attempt to keep the power down, the BTS will immediately begin to lower the power until mobiles start to complain. Each network manufacturer uses FER-based triggers when adjusting power. There are initial, minimum and maximum values. These set points are changed, when required, by the BSC.

Closed loop power control is broken down further into Inner and outer. Closed loop control is a more precise power control, which takes place when a mobile is in a dedicated mode (in a call or data session). Closed loop is broken down further into two types; inner and outer.

Forward closed inner loop is based on how the MS hears the BTS. The MS will tell the BTS to adjust its output power either up or down, depending on how strong or weak the received signal is. The MS can tell the BTS to go up or down 1 decibel 800 times per second.

Forward closed outer loop is also based on how well the MS hears the BTS. This time, the MS can have the BTS adjust the initial, minimum and maximum set points. These adjustments are temporary and are only in effect during that particular session. The set points would only need to be adjusted if the current values are not sufficient.

Reverse power control means that the power coming from the reverse/uplink is power controlled. As in forward power control, there are open, closed inner and closed outer.

Reverse open loop power control is used when an MS is accessing the network (R-ACH/R-EACH). There are two methods for this control. One is based on how well it hears the BTS (PICH - Pilot Channel). The MS will adjust its power up or down based on the received signal. For example, if it hears the BTS at a strong level, it will transmit at a low level. This is because it is assumed that a strongly received signal means that the MS is close to the tower. The opposite is also true. Based on this assumption, the MS will transmit at that power level. If it doesn't get a response from the BTS, it will increase its power and try again.

The other method is used when an MS has already registered with the network and has read the access parameter message. This message tells the MS what power level to use during access. It includes specific information, such as Initial power, nominal power and power steps.

Closed inner loop is used during dedicated sessions. It is the BTS who adjusts the mobile station's power. The BTS can tell the MS to increase or decrease its power 1 dB 800 times a second.

Closed outer loop is unique in that it is the BSC who plays a role. If the BSC has trouble hearing the MS (through the BTS of course), it can re-adjust the set points.

Rake Receiver

Rake receivers are utilized by both the BTS and MS to resolve multi-path issues while reducing fading in the operating environment. The Rake receiver is also used to manage soft handoffs. The RAKE receiver has multiple correlators and one searcher. The searcher is continuously looking for the best pilot channel quality while the correlators can be assigned to a single PN Offset or multiple PN Offsets to recover particular Walsh codes. The correlators can be targeted on delayed multi-path reflections or different BTSs when transitioning between BTSs during a soft hand-off.

Control Signaling

Control signaling is the process of transferring control information such as address, call supervision and other connection information between communication equipment and other equipment or systems. For the CDMA system, control channel signaling occurs on the separate pilot, synchronization, paging and access control channels before calls are setup. When a call is in progress, the CDMA channel either replaces the speech information (blank and burst) or mixes in the control information (dim and burst) with low-speed speech frames.

Blank and Burst Signaling

Blank and burst signaling is a process of sending control messages between telecommunications devices where control data temporarily mutes voice or user data and replaces it with a control message. This is also known as in-band signaling.

Blank and burst signaling on the CDMA radio channel is the process of replacing a 20 msec speech frame with a control message. Because a 20 msec time period is relatively short, and the audio information in one 20 msec speech frame is likely to be the same as the previous speech frame, the speech coder may replace the lost speech information with the previous speech frame instead of muting the audio signal.

Figure 1.6 shows the basic process of blank and burst signaling used in the CDMA system. This diagram shows that a control message replaces a frame of speech information. The information that is replaced is discarded. This example shows that the speech coder may replace the momentary loss of speech information with the previous speech frame as the amount of change in audio between 20 msec speech frames is relatively small.

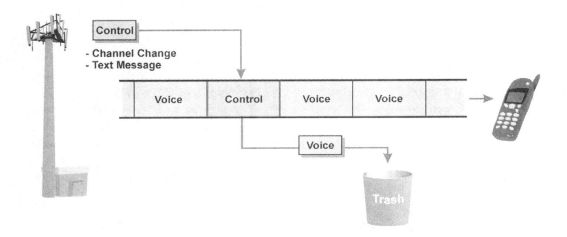

Figure 1.6, CDMA Blank and Burst Signaling

Dim and Burst Signaling

Dim and Burst signaling is the process of inserting (merging) control messages during periods of low speech or data activity. During dim and burst signaling, some of the traffic channel bits are used for control and some are used for audio or user data. Because the control message can only use some of the bits, it usually takes several frames to send a control message. The number of frames that are used to send a control message will vary based on the speech activity level or user data transmission rate.

To indicate that a dim and burst control message is mixed in with audio signals, a mixed mode flag bit is used (set) at the beginning of each frame that contains a control message. This allows the mobile device to decode the 20 msec frames as a control message instead of as a speech frame.

Figure 1.7 shows how CDMA dim and burst signaling may occur. This diagram shows that some of the 20 msec frame portion is used for audio and some of the 20 msec frame is used for the control message.

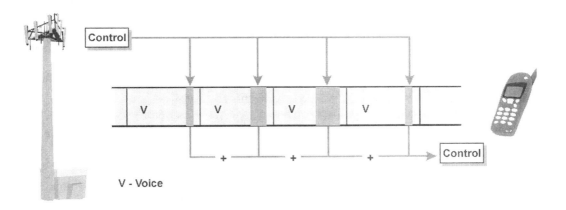

Figure 1.7, CDMA Dim and Burst Signaling

CDMA and Timing

CDMA is a synchronous network and uses synchronous time to sync all BTSs to network time. Most networks use GPS time plus UTC (Universal Time Coordinated) minus leap seconds.

CDMA Codes

The use of codes in CDMA can be misleading. The logical channels or data sets are the small pieces of information which, when combined with user specific information, will uniquely establish communications between the user and the network. Some of the information is referred to as codes. These codes are what set the data stream apart from other data streams being sent on the same frequency. In CDMA, there are short codes, long codes and Walsh codes. Some of these codes carry voice or data, while others provide routine signaling that keeps the call established while being mobile. Some of the logical channels are only used on the forward side, while others are only used on the reverse side. Finally, there are codes that work in concert with each other on both the forward and the reverse.

On the forward side, there are the short codes, the long codes and most Walsh codes. On the reverse side, there is the Public Long Code Mask and most Walsh codes.

The short code is the same code used by every sector of every BTS in the network. The purpose of the short code is to identify the particular sector on the forward link. Sector identification is achieved by shifting the start point of the 32,768 bit code by increments of 64 bits. This results in the possibility of 512 different Pseudo Noise Offsets (PN Offset) available per cdma_freq. The short code repeats every 26.666 milliseconds or 75 times every two seconds. The MS uses the PN Offset to first identify the sector, then to measure the strength and quality of the signal for re-selections and hand-offs.

In an effort to shorten the search time for the MS to find and acquire a signal, CDMA utilizes PN_Inc or PN Offset Increment values. Instead of searching all 512 possible values, the MS will search based on the increment pattern being used by the network. Typically, the PN_Inc is 3, 4, or 6. The network uses PN 3, 6, 9, 12, 15, 18, 21, or any of the 512 possible that are divisible by 3.

The long code is 4.4 trillion chips long, and repeats once every 41.2 days. The long code is used on the forward link to scramble the data being sent, while on the reverse side it uniquely identifies the MS from every other MS on that cdma_freq.

On the forward side the long code is the same for all sectors. The network will only send the state of the long code to the MS. This will allow the MS to determine where, within the 4.4 trillion chips, to find the part of the code the network is trying to correctly apply scrambling.

On the reverse side of the long code, the MS' ESN is applied to the long code state and then run through the long code generator to uniquely identify which MS the reverse transmission belongs to, as every MS is on the same cdma_freq.

There are many different Wash Codes (Logical Channels) used in CDMA. In IS-95 there are 64 (0-63) Walsh Codes available, while in CDMA 2000 there are 128 (0-127) Walsh Codes available.

Due to Quasi Orthogonal Functioning (QOF), there is technically no limit to the number of Walsh codes available for use in CDMA 2000. QOF is simply stating that, after a certain point on the Walsh code tree, there is enough buffer to prevent previous codes that had been used earlier from interfering with the longer codes being used further in the Walsh Code tree.

Direct Sequence Spread Spectrum (DSSS) orthogonal code properties conclude that, since all users are assigned a code, there must be some way to tell them apart, and that they do not interfere with each other. In radio communications, signals are said to be orthogonal if they do not interfere with each other.

Walsh Codes and Data Rates versus Data Protection is the process of Low Spreading Factor versus High Spreading Factor. For example, WC 2 has a higher data rate, meaning more data can be assigned to this Walsh code. However, Walsh Code 2 offers very low data protection. The other example WC 128 has a lower data rate but very high data protection. The idea is to use Walsh codes that meet in the middle for ideal data rates and data protection.

Below is a break out of what each Walsh code represents:

0 = Pilot Channel	1 = Paging Channel	2-7 = Paging Channel or Signaling/Traffic Channel
8-15 = Signaling/Traffic Channel	16 = Transmit Diversity Pilot Channel	17-31 = Signaling/Traffic Channel
32 = Sync Channel	33+ = Signaling/Traffic Channel	48, 80, 112 = F-QPCH

Figure 1.8, CDMA Walsh Code breakout

Logical Channels
Forward Logical Channels

The Forward Logical Channels are divided into two sections; Signaling and User. The Signaling Channels are further divided into Dedicated and Common Channels.

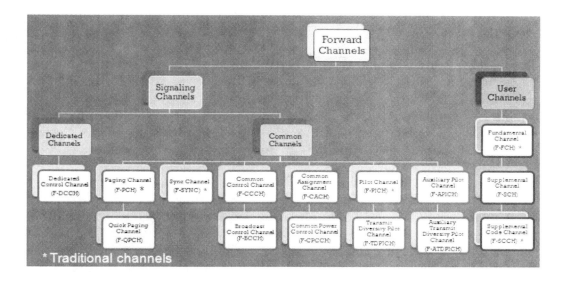

Figure 1.9, Forward Logical Channels

Signaling

Dedicated Channels

There is a single dedicated channel on the forward link of CDMA200. The Dedicated Control Channel (F-DCCH) is used for dedicated signaling and only allocated to one designated user. The F-DCCH may be used for short, low-rate user data, such as SMS messages. It is also used in conjunction with the Supplemental Channel for signaling purposes.

Common Channels

The Paging Channel (F-PCH) (WC 1-7) is used to contact mobile stations for incoming calls and data.

The Quick Paging Channel (F-QPCH) (WC 48, 80, 112) works with a normal paging channel (F-PCH). The user looks at the paging indicator bit sent on the F-QPCH. If there is a page for the mobile, it will wake up and read its paging channel slot. Also, the F-QPCH maintains the broadcast and configuration change indicators. These inform the MS if there has been a change to either the broadcast parameters or the configuration parameters.

The Broadcast Control Channel (F-BCCH) is used to transmit broadcast messages to all, including system parameter messages, neighbor lists and access parameter messages. This information is sent on the paging channel (F-PCH) in IS-95. Some of the information that is sent over the BCCH includes the system parameters message. This is where the MS finds the Base Station ID, Registration Zone ID, various rules for registering to the system and, if broadcasted, the location of the BTS. The Access Parameters Message is carried on the F-BCCH. This allows the MS to know how many access channels there are. Included in the F_BCCH traffic is the neighbor list message. This message is a list of Psuedo Noise Offsets (PN_Offsets), which tells the MS what offsets to measure for cell re-selection or hand-off, depending on what state the MS is in.

The Common Assignment Channel (F-CACH) is used to quickly assign Reverse Common Control Channel (R-CCCH) resources to the different mobile stations.

The Common Power Control Channel (F-CPCCH) is used to power control the Reverse Common Control Channel (R-CCCH) and the Reverse Enhanced Access Channel (R-EACH).

The Pilot Channel (F-PICH) (WC 0) is a beacon used as a timing source in network acquisition and as a measurement channel during idle and dedicated modes. All pilot channels transmit same power levels in an area.

The Transmit Diversity Pilot Channel (F-TDPICH) (WC 16) works with the F-PICH to support transmit diversity on the forward link. It uses two transmit antennas for spatial diversity to help with fading.

The Auxiliary Pilot Channel (F-APICH) is required for forward link spot beam and antenna beam forming applications, and provides a phase reference for coherent demodulation of those forward link CDMA channels associated with the auxiliary pilot.

The Auxiliary Transmit Diversity Pilot Channel (F-ATDPICH) is used with the Auxiliary Pilot Channel for transmit diversity, which increases forward gain.

User Channels

The Forward Fundamental Channel (F-FCH) is a forward traffic channel which carries voice and low-rate data. The Forward Fundamental Channel can also carry power control information, but may use the F-DCCH for this.

The Forward Supplemental Channel (F-SCH) is used for high-rate packet data. It does not send signaling data, so is used in conjunction with an F-FCH or F-DCCH.

The Forward Supplemental Code Channel (F-SCCH) is used in IS-95B to send data at 9.6 or 14.4 kbps.

Reverse Logical Channels

The Reverse Logical Channels are divided into two sections; Signaling and User. The Signaling Channels are further divided into Dedicated and Common Channels.

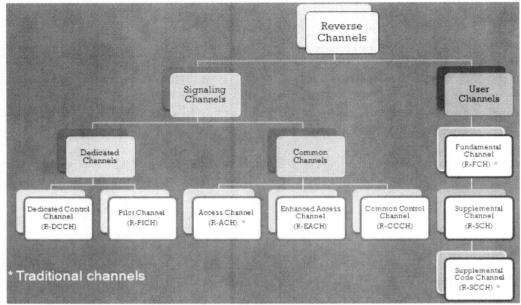

Figure 1.10, Reverse Logical Channels

Signaling

Dedicated Channels

The Dedicated Control Channel (R-DCCH) is used for dedicated signaling and only allocated to one designated user. It may be used for short low-rate user data, such as SMS messages. The R-DCCH is used in conjunction with the Supplemental Channel for signaling purposes.

Common Channels

The Reverse Pilot Channel (R-PICH) is an un-modulated, direct-sequence spread spectrum signal transmitted by a CDMA mobile station. A Reverse Pilot Channel provides a phase reference for coherent demodulation, and may provide a means for signal strength measurement. It differs from F-PICH in that it carries power control feedback so the base station can power control the forward link.

The Reverse Common Control Channel (R-CCCH) is a channel used for the transmission of digital control information from one or more mobile stations to a base station. It can be power controlled and may support Soft Handoff. The R-CCCH can be used when there has been no assignment of the R-DCCH or R-FCH.

The Reverse Access Channel (R-ACH) is used by the mobile to request network access. It is used for the transmission of short messages, such as signaling, response to pages, and call originations. It can also be used to transmit moderate-sized data packets. Each Access Channel is paired to a Paging Channel. Each Paging Channel can have up to 32 access channels.

The Reverse Enhanced Access Channel (R-EACH) is used by the mobile to request network access in CDMA 2000. The transmission is typically shorter than an R-ACH and, thus, the probability of collisions is smaller. It can be used for transmission of short messages, such as signaling, response to pages, and call originations. It can also be used to transmit moderate-sized data packets.

User Channels

The Reverse Fundamental Channel (R-FCH) is a reverse traffic channel which carries voice and low-rate data. The reverse Fundamental Channel can also carry power control information but may use the F-DCCH for this.

The Reverse Supplemental Channel (R-SCH) is used for high-rate packet data. It does not send signaling data. The R-SCH is used in conjunction with an R-FCH or R-DCCH.

The Reverse Supplemental Code Channel (R-SCCH) is used in IS-95B to send data at 9.6 or 14.4 kbps

A single Cdma_freq, as mentioned before, has the ability to handle up to 128 Walsh codes. Because of the noise involved, there would never be that many on any given Cdma_freq. There would be many transmitting various logical channels.

Figure 1.11 illustrates an example of various signaling and user channels broadcasted from a single channel. They include, for example, the Pilot channel, several fundamental channels, a supplemental channel, and dedicated control channels. All are broadcast simultaneously. Most of the signaling channels are constantly being broadcasted while user channels come and go based on user need.

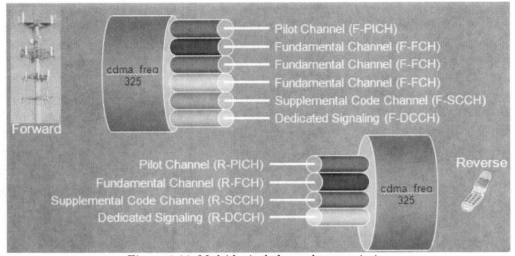

Figure 1.11, Multi logical channel transmission

Frequency Diversity

Frequency diversity is the process of receiving a radio signal or components of a radio signal on multiple channels (different frequencies) or over a wide radio channel (wide frequency band) to reduce the effects of radio signal distortions (such as signal fading) that occur on one frequency component, but do not occur (or do not occur as severe) on another frequency component.

Because the CDMA radio channel provides communication over a relatively wide 1.25 MHz radio channel (compared to the 30 kHz analog channel), it is less susceptible to signal fading. When radio signal fades occur (due to signal combining and canceling), they generally occur over a narrow frequency range. This means a signal fade only affects the reception of some of the chips that represent each bit of information that is transmitted. If a majority of the remaining chips can be successfully received, the result is the successful transfer of information, even in the presence of radio signal fades.

Figure 1.12 shows how a wideband radio channel offers the capability of frequency diversity. This example shows that only a portion of the wideband CDMA radio channel is affected by the radio signal fade. As a result, only a few of the chips are lost, and a majority of chips are successfully transmitted.

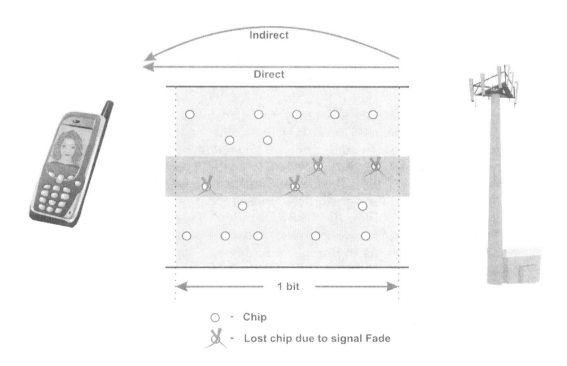

Figure 1.12, CDMA Frequency Diversity

Time Diversity

Time diversity is the process of sending the same signal or components of a signal through a communication channel where the same signal is received at different times. The reception of two or more of the same signal with time diversity may be used to compare, recover or add to the overall quality of the received signal.

The cause for time diversity may be naturally created or it may be self-induced. Delayed signals may be created by the reflection of the same signals by objects, such as buildings or mountains. When multiple signals are received that have taken different paths (direct and/or reflected), it is called multipath. In some cases, it is desirable to self-create multiple delayed signals in the transmitter. The creation of multiple delayed signals can be used to overcome the effects of signal fading. In either case, CDMA uses a receiver that is capable of decoding two or more signals that are delayed in time. This receiver is called a rake receiver.

Figure 1.13 shows how the CDMA system can use a rake receiver to combine multiple time delayed (multipath) signals to help produce a higher quality received signal. In this example, the same signal is received at the mobile device because one of the signals has been reflected off of a building during transmission. This reflected signal has to travel a longer distance and, as such, it is delayed slightly upon receipt by the mobile device. Because each signal is identified by a unique code, the receiver can separately decode each signal. The receiver can then select or combine the two (or more) reflected signals to produce a higher quality received signal.

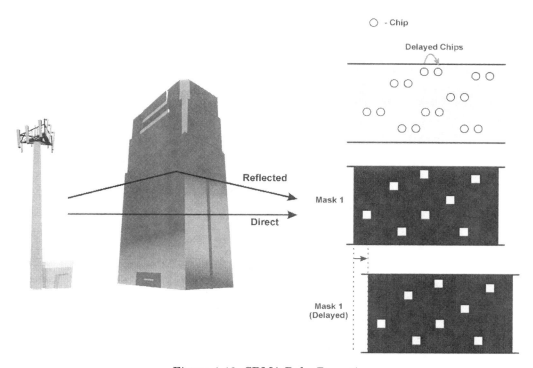

Figure 1.13, CDMA Rake Reception

Radio Coverage and Capacity Tradeoff

Radio coverage is a geographic area that receives a radio signal above a specified minimum level. The radio coverage area of a CDMA cell site can change based on the capacity used by customers. As more customers are added to a radio coverage area, the radio coverage area becomes smaller. This is known as Cell Breathing.

Cell breathing is an occurrence that takes place throughout the day. Based on subscriber usage, the tower's power will increase or decrease. For example, during peak hours (high network usage) the tower's power will be decreased. This reduces the amount of users per tower, which reduces noise. During non-peak hours there are less users utilizing the network resources. The tower's output is then increased.

Figure 1.14 shows how the CDMA system can tradeoff radio coverage for higher capacity. This diagram shows that, as more customers are added to the cell, the radio boundaries begin to contract into a smaller radio coverage area.

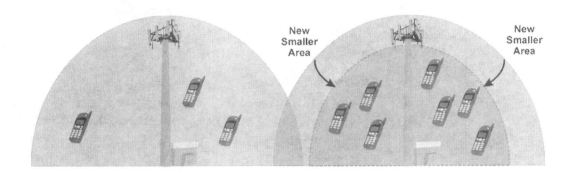

Figure 1.14, CDMA Radio Coverage and Capacity Tradeoff

Soft Capacity Limits

A capacity limit is the maximum amount of service (such as data transmission rate) or number of customers that a system can provide services to at a defined level of quality. The CDMA system has a soft capacity limit, as the system operator can dynamically change the defined level of service to change the maximum number of customers (capacity limit) who can obtain service from the system. This allows the service provider to temporarily increase the system capacity in exchange for a reduction in the quality of voice.

A CDMA service provider can increase the number of customers on a CDMA mobile communication system by reducing the audio quality through increased speech compression. This lowers the average data rate per user, reducing interference, and increasing the maximum number of users.

Figure 1.15 shows that a soft capacity limit allows for the gradual decay of voice quality in a communication system when additional users are added in a system. To provide service to more customers than are recommended by the capacity limit (over capacity) in a CDMA system, users in the system are provided with lower bit rates (higher speech compression). As a result of assigning lower bit rates to users as service demand increases, voice quality is traded off for increases in system capacity.

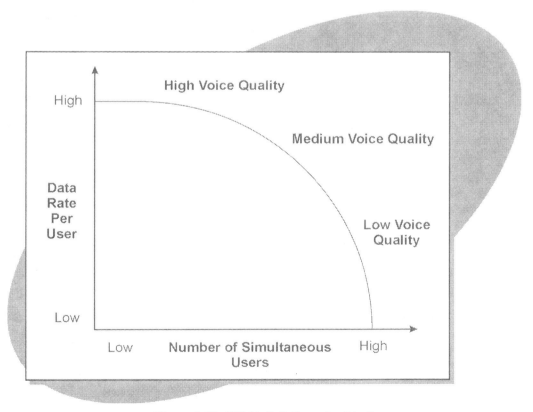

Figure 1.15, CDMA Soft Capacity Limit

Variable Rate Speech Coding

Digital speech compression (speech coding) is a process of analyzing and compressing a digitized audio signal, transmitting that compressed digital signal to another point, and decoding the compressed signal to recreate the original (or an approximation of the original) signal.

Figure 1.16 shows the basic digital speech compression process. The first step is to periodically sample the analog voice signal (20 msec) into pulse code modulated (PCM) digital form (usually 64 kbps). This digital signal is analyzed and characterized (e.g. volume, pitch) using a speech coder. The speech compression analysis usually removes redundancy in the digital signal, such as silence periods, and attempts to ignore patterns that are not characteristic of the human voice. In this example, this speech compression process uses pre-stored code book tables that allow the speech coder to transmit abbreviated codes that represent larger (probable) digital speech patterns. The result is a digital signal that represents the voice content, not a waveform. The end result is a compressed digital audio signal that is 8-13 kbps instead of the 64 kbps PCM digitized voice.

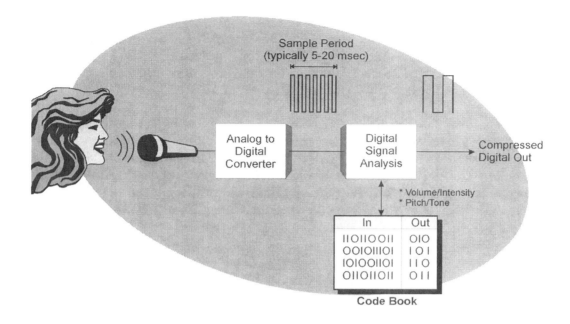

Figure 1.16, Speech Coding

Variable rate speech coding is a speech compression process that offers multiple speech coding rates. The use of variable compression rates allows a lower bit rate coding process (higher compression rates) to be used when system capacity is limited and more users need to be added to the system.

The variable rate speech coding typically begins by digitizing the audio signal to a 64 kbps audio signal. The speech coder then characterizes the speech signal into various components or parameters. By analyzing these components, the speech coder can select the data rate, which can be effectively used to recreate the audio signal. In periods of low speech activity, the number of bits needed to represent the audio signal may be low, while sections of dynamically changing audio signals may require more bits to represent the audio signal.

The CDMA system provides for the ability of the system to dynamically change the maximum data transmission rate that the speech coder may use to characterize (compress) the audio signal. While this maximum data rate limit may affect the quality of speech (less bits may mean lower audio quality in periods of high audio activity), the limit allows the CDMA system to control the maximum data transmission rate, and may allow the system to add additional users with a tradeoff of audio quality (called soft capacity).

Figure 1.17 shows the basic variable rate speech coding process. This diagram shows that a user is talking into a CDMA radio with a variable speech coder. In this example, the speech coder dynamically changes the data rate based on the speech activity level. During periods of high activity, the speech coder transmits at 9600 bps. At periods of low speech activity level, such as periods of silence, the speech coder transmits at 1200 bps.

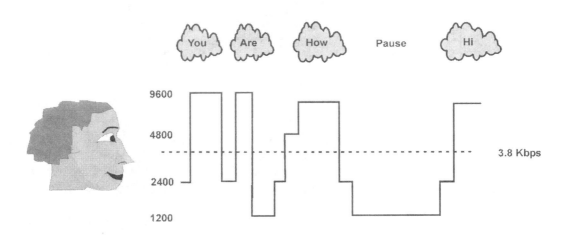

Figure 1.17, CDMA Variable Rate Speech Coding

Variable Spreading Rates

Variable spreading rates are the speeds at which the ratio of chips to information bits (spreading rate) can change in a spread spectrum communication system. The use of variable spreading rates allows the system to assign different data transmission rates on the same radio channel without having to change the chip rate.

The CDMA system uses fixed length spreading codes. To achieve variable data rates, the system either combines multiple coded channels (higher-data rates) or temporarily stops transmission during 1.25 msec power control groups (lower data rates).

Radio Configurations

Radio configuration is the combination of modulation, coding rate and spreading rate that is used on a radio channel. Two different radio configurations are used in the IS-95 CDMA system: RC1 and RC2. There are seven additional radio configurations used with cdma 2000. The radio configurations use different amounts of convolutional error protecting coding rates that produce different data rates.

Convolutional coding is an error correction process that uses the input information (data) to create a continuous flow of error protected bits. As these bits are inputted to the convolutional coder, an increased number of bits are produced. Convolutional coding is typically used in transmission systems that often experience burst errors such as wireless systems.

Radio Configuration 1 (RC1) uses convolutional coding rate 1/2, which provides a maximum data transmission rate of 9600 bps. Radio Configuration 2 (RC2) uses convolutional coding rate 1/2, which provides a maximum data transmission rate of 14400. Both RC1 and RC2 use BPSK during modulation. All other radio configurations primarily use QPSK modulation and a variety of coding rates.

Discontinuous Reception (DRx) Receiver Sleep Mode

Discontinuous reception (DRx) is a process of turning off a radio receiver when it does not expect to receive incoming messages. For DRx to operate, the system must coordinate with the mobile radio for the grouping of messages. The mobile radio will wake up during scheduled periods to look for messages.

The use of DRx allows the mobile device to reduce power or disconnect power to non-essential circuits during the off ("sleep") periods. The CDMA paging channel is divided into short 200 msec paging groups. The system can assign up to 640 paging groups. For normal mobile telephone paging (incoming calls), the system will assign 10 paging groups. One group represents the last digits of the mobile device telephone number (0 through 9). This results in the maximum delay of 2 seconds.

The maximum group sleep period is 2 minutes and 8 seconds. This allows the system to assign much longer paging groups to non-telephone devices such as data reporting devices. These data reporting devices may be power limited, such as earthquake seismic monitors or traffic monitoring sensors.

Figure 1.18 shows how paging groups can be used to provide for discontinuous reception capability. This diagram shows that the paging channels are divided into 200 msec groups, and that paging groups are typically associated with the last digital of the mobile device telephone number. This provides for 10 groups with a typical maximum delay of 2 seconds.

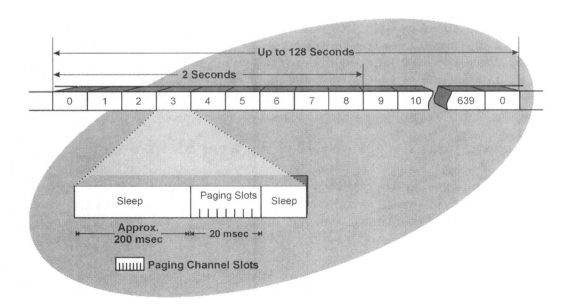

Figure 1.18, CDMA Discontinuous Reception

Discontinuous Transmission (DTX)

Discontinuous transmission is the ability of a mobile device or communications system to inhibit transmission when no activity, or reduced activity, is present on a communications channel. DTx is often used in mobile telephone systems to conserve battery life.

The CDMA system allows the mobile device to use DTx by intermittently stopping transmission during 1.25 msec time slots during periods of low data transmission activity. Each 20 msec frame is divided into sixteen 1.25 msec time slots. To ensure that the transmit bursts from multiple mobile devices do not occur at the same time, the transmit burst patterns (time slots on and time slots off) are varied (randomized). The previous speech frame determines the burst patterns.

The use of a variable rate voice coder makes the DTx operation possible, as the data rate varies based on voice activity. When the voice activity is low, the data rate is low, resulting in a limited number of transmit slots that are used for transmission during a 20 msec speech frame.

Figure 1.19 shows how the CDMA system can transmit in pseudo-random burst patterns with 1.25 msec slot time periods. This diagram shows that during the on burst, the transmitter is enabled to transmit at normal power (open loop + closed loop power control), and during the off burst, the transmitter is off. This example shows that the burst pattern is determined by the data transmission rate (9600 bps -vs- 4800 bps), and that the burst pattern code occurs over a 20 msec frame interval. The use burst transmission only occurs in the reverse direction. In the forward direction, the transmitted data is repeated during periods of low speed data instead of turning off the transmitter.

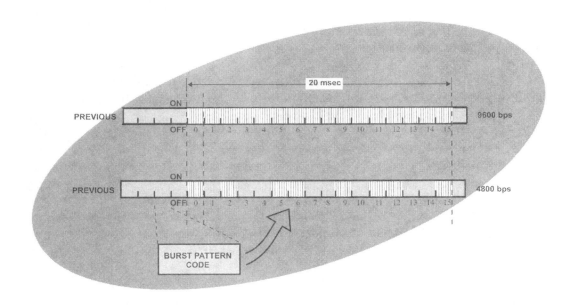

Figure 1.19, CDMA Transmit Power Bursts

CDMA Network Interaction

Basic interaction between the MS and the network can take place without any human intervention. As a matter of fact, it is typically done this way. Some things to consider include the various mobile transitions states, acquiring the system, registration and types, authentication, mobile terminate and originated calls, and handoffs.

Mobile Transition States

There are four transition states that any mobile station will be in at any given time, which include MS initialization state, MS idle state, system access state, and MS control on the traffic channel state.

The MS initialization state occurs when the MS is acquiring the CDMA network. It includes searching for CDMA energy in the bands, measuring pilot channels, reading the sync channel and syncing to the network.

The MS idle state occurs when the MS has already registered with the network. It is monitoring the various channels and measuring the environment for the purpose of reselection. It also officially includes the beginning of a mobile originated call and the period during which the MS begins to transmit a message.

The system access state occurs when the MS is communicating with the network, but is not in an actual traffic channel. This can be anything from responding to pages to conducting the registration process.

The MS control on the traffic channel state occurs when, as its name implies, the MS is actually in a traffic channel.

Initial Registration Process

When an MS acquires a CDMA system, or performs initial registration, it first performs a frequency scan based on the preferred bands for that MS. Its first goal is to find a cdma_freq that is on its history list and/or on the preferred channel list. When it does find CDMA energy, it will check it against its Preferred Roaming List (PRL). This list tells the MS if that SID is a home or roaming SID. Once a cdma_freq is found in the area, it will initially be used to measure all Pilot channels (PICH) it can read and determine the strongest one. It now reads the Sync channel to receive the necessary data for follow-on decoding. Next, it decodes (reads) the paging channel and reads all of the broadcast information coming from the chosen sector. This includes a channel list and/or extended channel list. Based on this list, it will perform hashing to determine the cdma_freq it will use to communicate with the network from that point on.

The MS will use RAKE fingers to decode and read the sync channel. The sync channel is always coded with Walsh Code 32, regardless of global location. The long code state includes crucial information, which will allow the MS to decode all of the information coming off of the tower.

After the sync channel is read, the MS reads the paging channel. Here, it further reads the messages required to further communicate with the network. The configuration messages include the channel lists, neighbor lists, parameter message and access message. These messages are sent across the Common Control Channel (CCCH) and Broadcast Control Channel (BCCH) in cdma2000.

There are two types of channel lists broadcasted in CDMA. The channel list is used for CDMAOne while the Extended Channel List is used in cdma2000. EVDO data is typically sent on a different channel and, thus, uses a separate channel list from standard 1X data.

Hashing in CDMA is used to load balance the network. When the network is setup as a multi-carrier network, meaning that it is broadcasting more than one cdma_freq from a sector, each MS runs their ESN or MIN, plus the number of available channels through the hashing function. The result is a slot position, which corresponds to an available cdma_freq.

Besides the hashing for load leveling, the network has another option. It may use what is called Global Service Redirection. Whether the network uses one or the other is based on what is sent across the parameter message. It is setup on a per sector basis, which allows the network operator to set different traffic distribution schemes throughout the multi carrier coverage area, based on actual traffic patterns.

If Global redirect is used, the MS will use the last digit of its ESN to decide which cdma_freq to use. This is known as the Access Overload Class.

Registration Types

Registration is crucial in any cellular network. It lets the network know that the MS is on and able to accept incoming data. In CDMA, there are several conditions that require the MS to register. Powering on the MS is just one of these. The types of registration required on the network are sent out across the system parameter message.

Registration types include zone based, timer based, parameter change, distance based, power-up, power-down, ordered, implicit, TCH, foreign SID, and foreign NID.

Power-up and power-down registrations are simple. The MS will register with the network when it does these types of registrations.

Timer-based registration is similar to the T-3212 in GSM. When the timer expires, the MS will re-register. When the MS comes out of idle state, the timer is reset and begins again.

Distance-based registration requires the MS to re-register when a distance threshold is met. It is based on the last tower it registered with and the current tower it is using.

Zone-based registration is the most commonly used type of registration for CDMA networks. Whenever an MS crosses over to another Reg_Zone, it is required to re-register.

Parameter-change registration is primarily based on when an MS uses a tower with an NID that is different from the one it is coming from. Also, if the capabilities of the MS are changed, the MS must re-register and include the changes. The phone capabilities are identified as a station class mark.

Ordered registration is just that. When an MS is ordered to do so, it will re-register.

Implicit registration requires that any time an MS sends an origination message, or replies to a page, it is implied that the MS is still turned on and active.

The traffic channel registration is about the same as Implicit. If the MS is put in to a TCH (FCH, SCH, SCCH), it is still turned on and active.

When a mobile roams into a foreign SID (an SID that is not in the preferred roaming list), it is still required to register. Even though it may not have full access to the network, it still can be used for emergency calls.

When a mobile roams into a foreign NID (an NID that is not in the preferred roaming list), it is still required to register. This will show up on the MS as "roaming." Based on the network, it may or may not have full access to network resources.

Authentication Process

The authentication process is used to verify that an MS is valid and is authorized to utilize the network. The process includes eight basic steps:

1. The HLR/AC will combine the mobile's A-Key and ESN and locally produced RANDSSD through a CAVE algorithm

2. The RANDSSD is also forwarded to the MS where it performs the same function

3. The CAVE algorithm produces two items; the SSD_A and SSD_B

4. The SSD_A is sent to the MSC where it combines the SSD_A, ESN, MIN, and a locally produced RAND through the CAVE algorithm

5. The RAND is sent to the MS where it performs the same process

6. The CAVE algorithm produces an Authentication Signature (AUTHR)

7. The MS sends the AUTHR to the MSC where it is compared with the AUTHR produced locally

8. If they match, the MS is authenticated

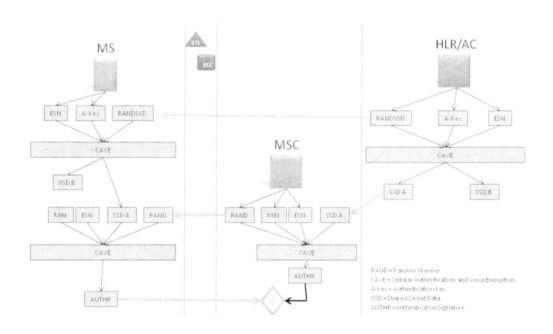

Figure 1.20, CDMA Authentication Process

Mobile Terminated Call

A Mobile Terminated Call (MTC) is when an MS receives a call. While the MS is idle it is reading the F-PCH, F-BCCH, and F-CCCH for any parameter changes or for any pages for it. When an incoming call is received, a general paging message is sent to the MS. The MS responds with a paging response on either an R-EACH or on an R-ACH. The network will then set up a traffic channel (FCH) and send a channel assignment message.

The next step is for the MS and BTS to notice and confirm each other's presence by exchanging acknowledgement messages on the traffic channel. The BTS and MS then negotiate on what type of call this will be. Next, the mobile is told to ring and given a "caller line ID" to display.

Mobile Originated Call

A mobile originated call (MOC) is when the MS makes the call. The user enters the digits and hits send. An origination message is sent on the R-EACH or R-ACH. The BTS sends an acknowledgement message on either the F-PCH or F-CCCH.

The network arranges the resources for the call and starts transmitting on the traffic channel. The network notifies the MS in a channel assignment message on either F-PCH or F-CCCH. The MS then goes to the traffic channel.

The MS then sends a TCH preamble on the R-PICH, which the BTS uses for power control. The MS and BTS exchange acknowledgement messages on the traffic channel (FCH). They negotiate the type of call and then the audio circuit is complete. At this point, the caller hears ringing in the phone.

Ending a Call

A normal call continues until one of the parties hangs up. This action sends a release order as "Normal Release." The other side of the call then sends a release order as "no reason given." If the normal release is visible, the call ends normally.

At the conclusion of the call, the mobile will reacquire the system. It will again search for the best pilot channel, read the Sync channel, and then monitor the paging channel.

Handoff Types

There are two handoff categories; Hard and soft. Hard handoffs involve the MS as it drops one physical channel and retunes to another physical channel. This means that it must break the link prior to making the new link. This is known as Break-Before-Make. All GSM handsets use this type due to every sector using a different physical channel. CDMA can also use this type of handoff if it is selecting a new channel, or if it moves into a new area with different channels. Typically, CDMA uses what is known as soft handoffs.

Since CDMA broadcasts the same frequency (cdma_freq) throughout an area, the MS performs soft handoffs where it establishes a connection with the new sector before breaking the connection with the losing sector. This is known as Make-Before-Break. There are actually three main types of soft handoffs; soft, softer, and soft-softer. A soft handoff is done from one sector of a BTS to another sector of another BTS. A softer handoff occurs when the MS hands off from one sector to another sector on the same BTS. The last type is a soft-softer, which is a combination of the first two. Here, the MS is handed off between combinations of two sectors of one BTS to another sector of a different BTS.

When an MS is idle and changes serving cells (sectors), it is known as an idle mode handoff or a PN re-selection. When the MS is idle, it continuously measures all of the pilot channels it can read from the neighbor list. When another pilot channel (PN offset) becomes stronger by at least 3 dB, it will start using the new PN offset. If the new PN offset requires registration, it will at that time. If not, it doesn't need to interact with network.

When the MS is in idle mode, it monitors a set of pilot channels. These are known as idle mode pilot sets. There are four sets while in idle mode, which include Active, Neighbor, Private Neighbor, and Remaining. The active set is the pilot to which the mobile is currently attached. The Neighbor set includes pilots (up to 40) that are transmitted by BTS via a Neighbor List message. The Private Neighbor set includes pilots (up to 40) that are transmitted by BTS via a Private Neighbor List message. The Remaining set includes all remaining possible pilots in the system that may be used.

Handoffs occur while the MS is in a call. This is known as an in-call handoff, and it is controlled by the network. The MS assists the network as it sends reports on how well it sees the pilot channels it is told to monitor. How well it sees a pilot is based on an Ec/Io value. The network will send the MS a handoff direction message when it wants it to either start using a pilot or to stop using a pilot.

An Ec/Io value is based on how "clean" (less noisy) the pilot channel is. This value foretells the readability of the associated traffic channels. The value guides the network when making soft handoff decisions. The value is a ratio of good to bad energy seen by the search correlator of the RAKE receiver. In other words, it is a ratio of the total energy received from the pilot channel alone divided by the total power received from that sector. Strong RF and noise from other sectors can also degrade the Ec/Io values.

The pilot sets for in-call handoffs are different than idle mode. They consist of four categories; Active, Candidate, Neighbor, and Remaining. The Active set includes pilots (up to 6) associated with the Forward Traffic Channels assigned to the MS and in use. The Candidate set includes the pilots (up to 10) that the MS has reported to be of sufficient signal strength (exceeds T_ADD). The Neighbor set includes pilots (up to 40) that are not currently on the Active and Candidate List, but are transmitted by BTS via Extended Neighbor List message. The Remaining set includes all remaining possible pilots in the system that may be used.

Handoff parameters determine when a PN offset is added and removed throughout the pilot sets previously mentioned. The first is called T_ADD. This is known as pilot detection threshold and is a value of Ec/Io at which a pilot is moved into the candidate set usually from the neighbor set.

T_COMP, or comparison threshold, is the threshold value of the ratios of Ec/Io at which a pilot in the Candidate Set may be moved to the Active Set.

T_DROP (Pilot Drop Threshold) is the threshold value of Ec/Io at which a counter is started to determine if that pilot should be removed from the Active Set.

T_TDROP (Drop Timer Threshold) is the drop-timer threshold interval which is the threshold of time beyond which a pilot with Ec/Io below T_DROP is removed from the Active Set.

The measurements that the MS sends are known as Pilot Strength Measurement Messages. They are sent:

When a pilot in the MS's neighbor list exceeds T_ADD
When a candidate set pilot exceeds T_COMP
When an active set pilot drops below T_DROP for T_TDROP

References

1. Lawrence Harte, Bruce Bromley, Mike Davis (2009). Introduction to GSM, Physical channels, logical channels, network functions, and operations. Fuquay-Varina: Althos publishing

2. James Harry Green (2000). The Irwin Handbook of Telecommunications. McGraw-Hill

3. Samuel C. Yang (2004). 3G CDMA2000 Wireless System Engineering. Boston: Artech House, Inc.

4. 2,3G. (2011). 3GPP2 Specifications. Retrieved from http://www.3gpp2.org/Public_html/specs/alltsgscfm.cfm

5. 3GPP. (2011). Index of /ftp/Specs/latest/R1999. Retrieved from http://www.3gpp.org/ftp/Specs/latest/R1999/

Index

Index

Voice Coding, 48-50
Wideband, 2, 7, 43

CPSIA information can be obtained at www.ICGtesting.com
Printed in the USA
BVOW06s0005240713

326736BV00003B/8/P